Mobile Communications Basics

FIRST EDITION

BY PRASUN BARUA

ABOUT

Welcome to Mobile Communications Basics! This is a nonfiction science book which contains various topics on basics of mobile communications. Mobile communication refers to the application of technology that enables remote communication without the need for a physical connection (wires or cables). Our lives are made simpler by mobile communication, which also saves time and effort. A mobile phone, also known as a cell phone, hand phone, or mobile cellular network, is one type of mobile communication (wireless communication). It is an electric device that transmits and receives data over a cellular network of base stations known as a cell site using full duplex two-way radio. In Europe and other parts of the world, many mobile phone users use the GSM (Global System for Mobile communication) digital mobile network. The most popular of the three digital wireless telephony technologies—TDMA, GSM, and CDMA—GSM uses a variation of time division multiple access (TDMA) (CDMA). GSM converts data to an electronic form, compresses it, and sends it along with two other streams of user data, each in its own time slot, down a channel. It uses a variety of frequency bands. The mobile device itself, the base station subsystem (BSS), the network switching subsystem (NSS), and the operation and support subsystem are the four separate components that make up the GSM network and cooperate to operate as a whole (OSS). Hardware is used to connect the mobile device to the network. The network receives identification data about the mobile user from the subscriber identity module (SIM) card. In this book, concepts of mobile communications, GSM and its architecture, GSM radio network concepts and call processing in GSM are discussed. This is the first edition of the book. Thanks for reading the book.

CONTENTS

CHAPTER-1: CONCEPTS OF MOBILE COMMUNICATIONS

From early to modern ages, mankind has sought long-distance communication methods. For centuries, letters proved to be the most dependable method of communication. To transmit information faster, fire, flags, horns, and so on were used. Technical advancements in the nineteenth century facilitated long-distance communication, resulting in Telegraphy and, later, telephony. Both methods used wire lines. J.C.Maxwell established the electro-magnetic theory in 1873, which is still valid today. However, it would be several decades later (in 1895) that Marconi would put this theory to practical use by developing devices for wireless transmission of Morse signals (in 1895). The first wireless voice transmission occurred in 1906 (R. Fesseden), and one of the first radio broadcast transmissions 1909 occurred in New York.

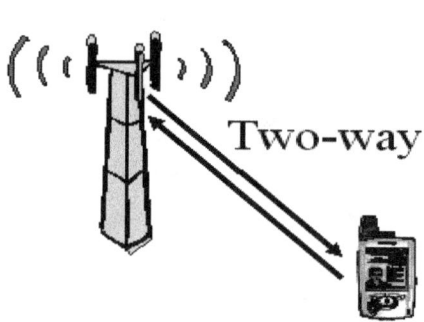

Radio broadcasting was the most economically successful wireless application in the first half of the twentieth century. There is only one transmitter, known as a radio station. The radio station transmits information such as news, music, and so on to the receiver equipment, or radio device. Simplex transmission is a type of one-way transmission. Transmission occurs only in one direction, from transmitter to receiver. This was the first fixed wireless transmission method. A technical solution that allows information to flow in both directions is required for conversation. Duplex transmission is the name given to this type of transmission. The walkie-talkie was already available in the early 1930s. This system already allowed user data to be transmitted in both directions, but there was a problem. The users were not permitted to transmit at the same time. In other words, you could only receive or transmit user data at any given time. As a result, this mode of transmission was commonly referred to as semi-duplex transmission. Telephony services necessitate technical solutions that give subscribers the impression that they can speak (transmit) and hear (receive) at the same time.

Full duplex transmission is the term used to describe this type of transmission solution. Mobile Telephony was created as a result of limited mobility and duplex transmission. The first commercial wireless car phone telephone service began in St. Louis, Missouri in the late 1940s (USA). It was a car phone service because mobile phone equipment was bulky and heavy at the time. Actually, it took up the entire back of the car during startup. However, it was a true full duplex transmission solution.

Several vehicle radio systems were installed in Europe in the 1950s as well. These are known as single cell systems. User data is transmitted between the mobile phone and the base station (BS). A base station sends and receives data from users. A mobile phone is only responsible for its user's data transmission and reception, whereas a base station can handle the calls of multiple subscribers at the same time.

The transmission of user data from the base station to the mobile phone is referred to as downlink (DL), while the transmission from the mobile phone to the base station is referred to as uplink (UL). The base station supply area, also known as cell, is the area where wireless transmission between mobile phones and the base station can take place.

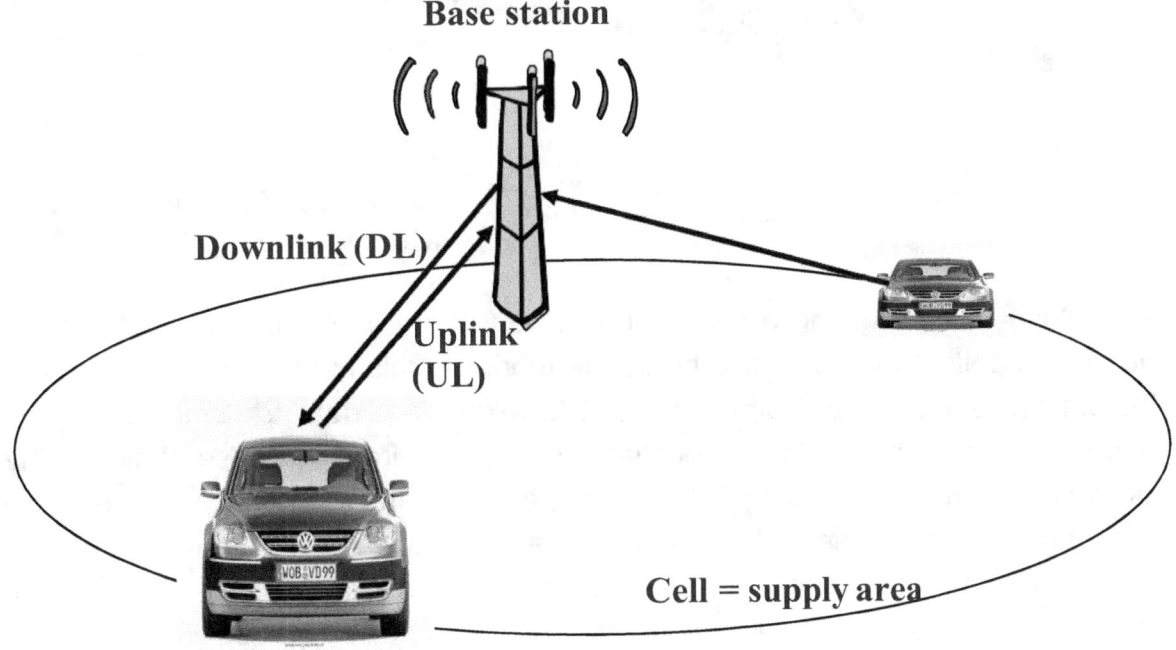

Single cell systems have very few applications. The quality of the radio link decreases as the subscriber gets further away from the base station. Communication is no longer possible if the subscriber leaves the cell's supply area. In other words, the cell was the only place where the mobile communication service was available. Multi-cellular systems were developed to get around this restriction. There are many cells in a cellular mobile communication system, and some of them may overlap. By doing this, the mobile communication service can be provided to an entire region.

But what happens when a subscriber switches from one cell to another during a call? It would be extremely inconvenient if the call was dropped. If a subscriber leaves a cell while simultaneously entering another, the system makes new radio resources available in the neighboring cell, and the call is transferred from one cell to the next. Service continuity is ensured as a result, even if the subscriber moves. This is known as handover (HO).

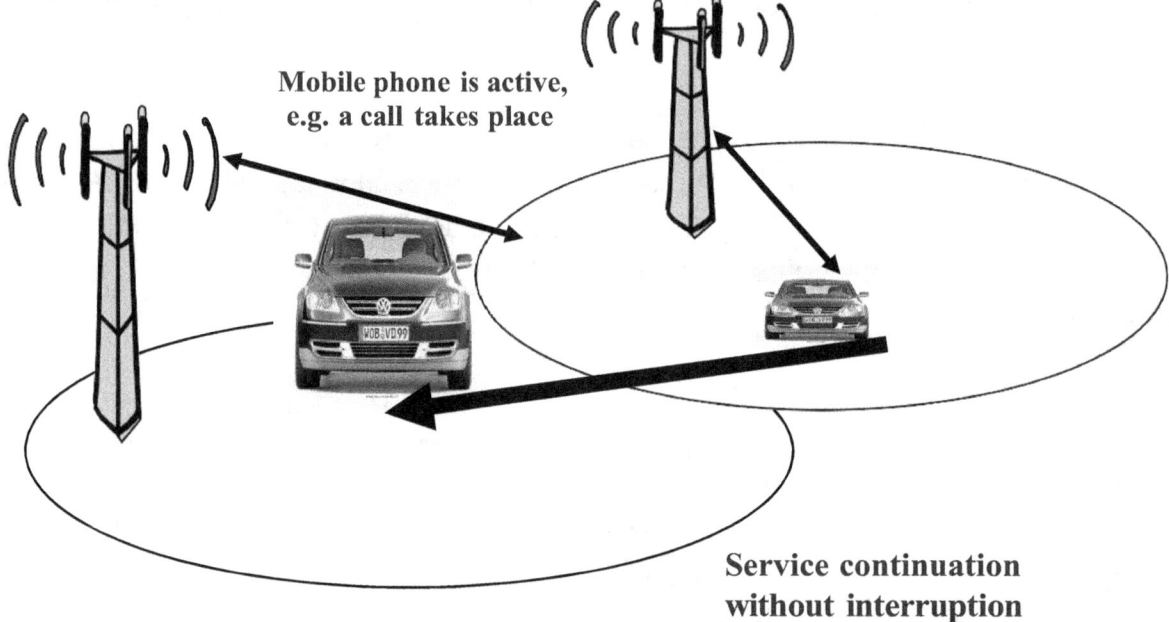

A handover occurs during a call, that is, when the phone is in active (dedicated) mode. A mobile phone can also be in idle mode. In this case, the phone is turned on, but no resources are allocated to it to allow user data transmission. In this mode, the mobile phone is still receiving information from the base station. Why? Assume there is an incoming call on this phone. In the cell, the mobile phone is then paged. This implies that the phone receives notification that a mobile call has been terminated. A cellular system can have hundreds of cells. If the mobile network is unsure which cell the phone is in, it must be paged in all of them. Paging is done in small parts rather than to a group of cells in a mobile network to reduce network load. Location area refers to the group cells in administrative units in an operation (LA). A mobile phone can only be paged in one location area at a time. The GSM (Global System for Mobile communication) system employs the LA to locate a subscriber in an active state. GSM is a digital mobile network that is widely used by mobile phone users in Europe and other parts of the world.

However, how does the cellular system know where the mobile phone is located? And how does the phone know this? System information is continuously transmitted in each cell. The location area information is included in the system information. The mobile phone is listening to this system information while in idle mode. If the user moves from one cell to the next and the new cell is in the same location area, the mobile device remains idle. If the new cell belongs to a new location area, the phone must be activated. It initiates network communication and sends data to the mobile network. This information is stored in databases within the mobile network, and if a mobile call is terminated, the network knows where to page the subscriber. Location Update Procedure refers to the process by which a mobile phone informs the network of its new location (LUP). The Mobile is registered at the VLR (Visitor Location Register) associated with the Mobile Switching Network.

CHAPTER-2: GSM AND ITS ARCHITECTURE

What is GSM?

GSM stands for Global System for Mobile Communications. It is an international standard for digital cellular communications. It is a digital mobile network popular among mobile phone users in Europe and other parts of the world. GSM is the most widely used of the three digital wireless telephony technologies: time division multiple access (TDMA), GSM, and code-division multiple access (CDMA) (CDMA). GSM digitizes and compresses data before transmitting it along with two other streams of user data, each in its own time slot, down a channel. It operates in various frequency bands.

GSM Frequency Bands

From MS to BTS or from BTS to MS the user information (Voice & Data) and the control information (signaling between MS & GSM Network) for authentication, Location Update Procedure, call setup, disconnection etc, is transmitted through the air interface over the Radio Carrier frequencies. On the basis of Radio carrier frequency band used a number of GSM systems have been developed as tabulated below:

GSM System	Up-Link	Down-Link
GSM - 850	824 – 849 MHz	869-894 MHz
GSM - 900	890-915 MHz	890-915 MHz
GSM - 1800	1710-1785 MHz	1710-1785 MHz
GSM - 1900	1710-1785 MHz	1930-1990 MHz

Today, most of the mobile handsets support multiple bands as used in different countries. These are typically referred to as multi-band phones. Dual-band phones can cover GSM networks in pairs such as 900 and 1800 MHz frequencies or 850 and 1900. European tri-band phones typically cover the 900, 1800 and 1900 bands giving good coverage in Europe and allowing limited use in North America, while North American tri-band phones utilize 850, 1800 and 1900 for wide-spread North American service but limited world-wide use. A new addition has been the quad-band phone, supporting all four major GSM bands, allowing for global use.

GSM NETWORK STRUCTURE

Every telephone network needs a well-designed structure in order to route incoming called to the correct exchange and finally to the called subscriber. In a mobile network, this structure is of great importance because of the mobility of all its subscribers [1-4]. In the GSM system, the network is divided into the following partitioned areas.

- GSM service area;
- PLMN (Public Land Mobile Network) service area;
- MSC (Mobile Switching Center) service area;
- Location area;
- Cells.

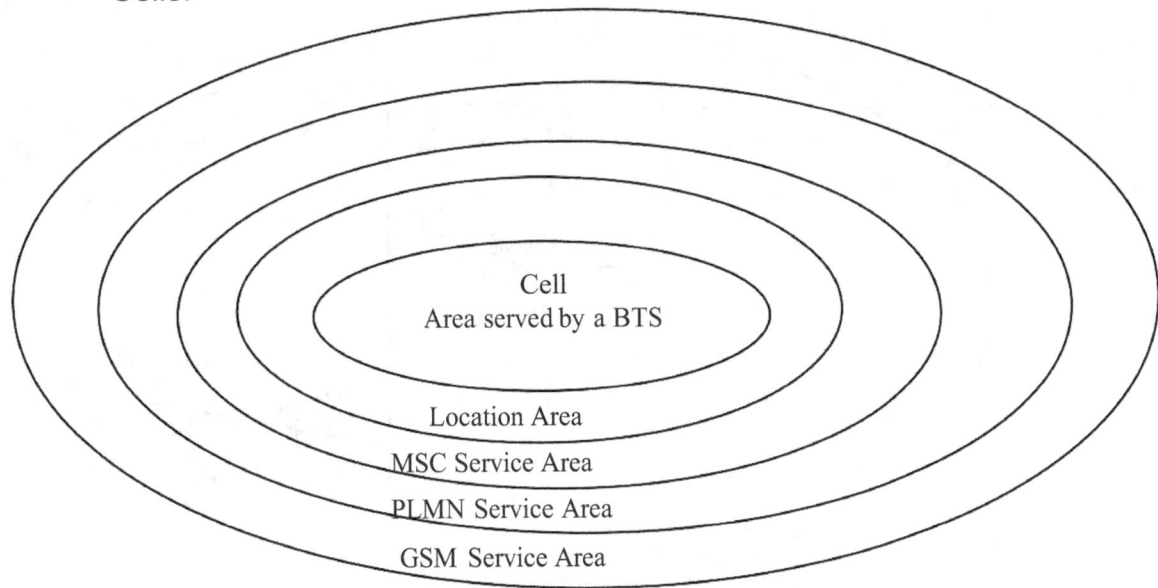

The GSM service is the total area served by the combination of all member countries where a mobile can be serviced. The next level is the PLMN service area. There can be several within a country, based on its size. The links between a GSM/PLMN network and other PSTN, ISDN, or PLMN network will be on the level of international or national transit exchange. Call connections between PLMNs, or to fixed networks, must be routed through certain designated MSCs called a gateway MSC. All incoming calls for a GSM/PLMN network will be routed to a gateway MSC. A gateway MSC works as an incoming transit exchange for the GSM/PLMN. In a GSM/PLMN network, all mobile-terminated calls will be routed to a gateway MSC. The gateway MSC contains the inter-working functions to make these connections. They also route incoming calls to the proper MSC within the network. The next level of division is the MSC service area. In one PLMN there can be several MSC service area. MSC has the role as a controller of calls within its jurisdiction. In order to route a call to a mobile subscriber, the path is through to the MSC in the MSC area where the subscriber is currently located. The mobile location can be uniquely identified since the MS is registered in a VLR, which is generally associated with an MSC.

The next division level is that of the LA's within a MSC combination. There are several LA's within one MSC combination. A LA is a part of the MSC/VLR service area in which a MS may move freely without updating location information to the associated MSC that control the LA. Within a LA a paging message is broadcast in order to find the called mobile subscriber. The LA can be identified by the system using the Location Area Identity (LAI).

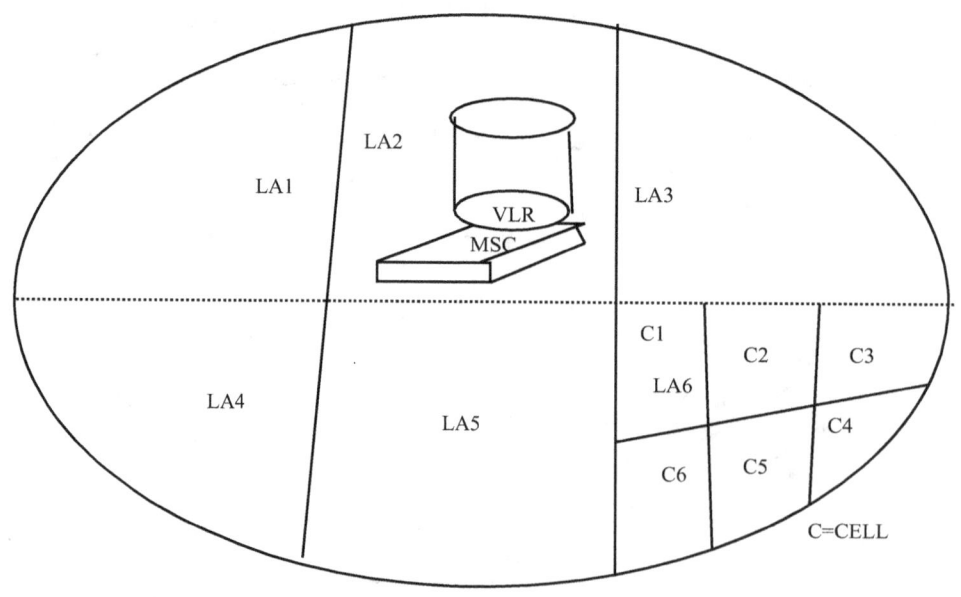

Lastly, a LA is divided into many cells. A cell is an identity served by one BTS. The MS distinguishes the cells by the Base Station Identification Code (BSIC) of each cell broadcast over the air.

MOBILE STATION

The wireless mobile telephone used by the subscriber is called the MS (Mobile Station). MS includes radio equipment and the man machine interface (MMI) that a user needs, in order to access the services provided by the GSM PLMN. MS can be installed in Vehicles or can be portable or hand-held. The MS may include provisions for data communication as well as voice.

Functions of MS

The primary functions of MS are to transmit and receive voice and data over the air interface of the GSM system. MS performs the signal processing function of digitizing, encoding, error protecting, encrypting, and modulating the transmitted signals. It also performs the inverse functions on the received signals from the BS.

In order to transmit voice and data signals, the mobile must be in synchronization with the network system so that the messages can be transmitted and received by the mobile at the correct instant. To achieve this, the MS automatically tunes and synchronizes to the frequency and time with the network.

The MS continuously monitors the power level and signal quality received on the downlink, by reading the error rate and strength of the signals received from its current BTS and the six surrounding BTSs. The MS sends this information to the BTS & BSC to facilitate the network to take decision on handover.

MS keeps the GSM network informed of its location during both national and international roaming, even when it is inactive. This enables the System to page it in its present LA.

Finally, the MS can store and display short alphanumeric messages on the liquid crystal display (LCD) that is used to show call dialing and status information. These messages are limited to 160 characters in length. (Some mobiles handsets are capable of taking more characters and then splicing into smaller parts while sending)

Power Levels

There are five different categories of mobile telephone units specified by the European GSM system: The 20-W and 8-W units (peak power) are either for vehicle-mounted or portable station use. The handsets normally used are of 2W.

The MS power is adjustable in steps from its normal value down up to 20mW. This is done automatically under remote control from the BTS, which monitors the received power and adjusts the MS transmitter to the minimum power setting necessary for reliable transmission.

Subscriber Identity Module (SIM) Card

The SIM is a removable smart card bearing a unique identification number. At the very beginning of the service, GSM subscribers are provided with a SIM card. The subscriber is identified in the system when the user inserts the SIM card in the mobile equipment. This provides an enormous amount of flexibility to the subscribers since they can now use the SIM in the mobile equipments, as the SIM card is portable between Mobile Equipment (ME) units. The user only needs to take his smart card on a trip. He can then rent a ME unit at the destination, even in another country, and insert his own SIM. Any calls he makes will be charged to his home GSM account. Also, the GSM system will be able to reach him at the ME unit he is currently using.

The SIM contains an integrated circuit chip with a microprocessor, random access memory (RAM), and read only memory (ROM).

When a mobile subscriber wants to use the system, user mounts the SIM card and provides the Personal Identification Number (PIN), which is compared with a PIN stored within the SIM. If the user enters three incorrect PIN codes, the SIM is disabled. The PIN can also be permanently bypassed by the service provider, if requested by the subscriber. Disabling the PIN code simplifies the call setup but reduces the protection of the user's account in the event of a stolen SIM card.

Each MS is identified by a unique identification number named as International Identification Number (IMEI), which is permanently stored in the mobile unit. Upon request, the MS sends this number over the signaling channel to the MSC. The IMEI can also be used to identify mobile units those are reported stolen or operating incorrectly.

Just as the IMEI identifies the mobile equipment, mobile subscribers are identified by their internationally unique numbers named as International Mobile Subscriber Identity (IMSI). Different subscriber identities are used in different phases of call setup. The IMSI is the primary function to identify the subscriber within the mobile network and is permanently assigned to him. The Mobile Station International Subscriber Directory Network Number (MSISDN) is the number that the calling party dials in order to reach the subscriber. It is used by the landline or other network to route calls toward an appropriate MSC.

International Mobile Equipment Identity (IMEI)

The IMEI is the unique identity of the equipment used by a subscriber and is used to determine authorized (white), unauthorized (black), and malfunctioning (gray) GSM hardware. In conjunction with the IMSI, it is used to ensure that only authorized user is granted access to the system. An IMEI is never sent in cipher mode by MS.

International Mobile Subscriber Identity (IMSI).

An IMSI is assigned to each authorized GSM user. It consists of a mobile country code (MSC), mobile network code (MNC), and a PLMN unique mobile subscriber identification number (MSIN). The IMSI is not hardware-specific. Instead, it is maintained on a SC by an authorized subscriber and is the only absolute identity that a subscriber has within the GSM system. The IMSI consists of the MCC followed by the NMSI and shall not exceed 15 digits.

Mobile Station International Subscriber Directory Number (MSISDN)

The MSISDN number is composed of the country code (CC) followed by the National Significant Number (N (S) N), which shall not exceed 15 digits. In order to obtain a mobile subscriber in another country, the MS should dial an international prefix before the MSISDN. (Example: 919412024567)

Network Details:

(1) Base Station Subsystem (BSS) :

BTS	-	Base Transrecieve Station
BSC	-	Base Station Controller
MS	-	Mobile Station

(2) Network Switching Subsystem (NSS):

MSC	-	Mobile Switching Centre
VLR	-	Visitor Location Register
HLR	-	Home Location Registered
AuC	-	Authentication Centre
EIR	-	Equipment Identity Register

(3) OMC   -   Operation & Maintenance Centre

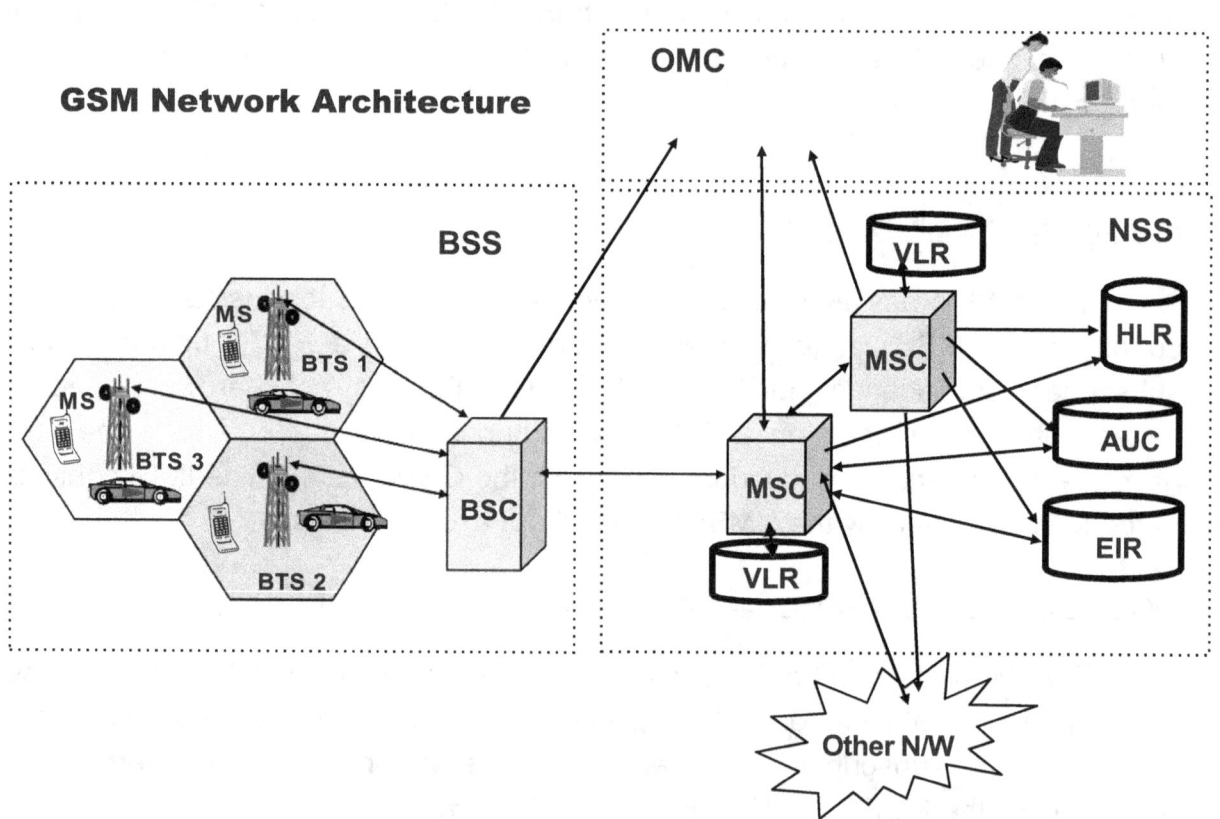

GSM Network Architecture

(4) Important Intefaces:

Interface		Interfacing Nodes
Um	-	MS and BTS
Abis	-	BTS and BSC
A	-	BSC and MSC

BASE STATION SUBSYSTEM (BSS)

The BSS is a set of Base Station equipments, such as BTS & BSC, responsible for communicating with MSs in a certain area. The BSS includes two types of machines: the BTS in contact with the MSs through the radio interface and the BSC, the latter being in contact with the MSC. BSS is the system interfaced with MSC. BSC is the nodal unit in the BSS. A BSC is a network component in the PLMN that functions for the control of one or more BTS. It is a functional entity that handles common control functions within a BTS. Typically, one BSC can control more than one BS or BTS. The radio equipment of a BSS may be composed of one or more cells. A BSS may consist of one or more BTS. The mobile function is divided basically between transmission equipment, the BTS, and managing equipment at the BSC. A BTS comprises of radio transmission and reception devices, including the antennas, and also all the signal processing specific to the radio interface. A single transceiver within BTS supports eight basic radio channels.

An important component of the BSS that is considered in the GSM architecture is the Transcoder/Rate Adapter Unit (TRAU). The TRAU is the equipment in which coding and decoding is carried out as well as rate adaptation in case of data. Although the specifications consider the TRAU as a sub-part of the BTS, it can be sited away from the BTS (at BSC), and even between the BSC and the MSC.

The interface between the MSC and the BSS is a standardized SS7 interface (A-interface) that is fully defined in the GSM recommendations. This allows the system operator to purchase switching equipment from one supplier and radio equipment and the controller from another. The interface between the BSC and a remote BTS is a standard interface known as the A-bis. In splitting the BSS functions between BTS and BSC, the main principle was that only such functions that had to reside close to the radio transmitters/receivers should be placed in BTS. This will also help reduce the complexity of the BTS.

Functions of BTS

A BTS is a network component that serves one cell and is controlled by a BSC. BTS is typically able to handle three to five radio carries, carrying between 24 to 40 simultaneous communications. Reducing the BTS volume is important to keeping down the cost of the cell sites. The primary responsibility of the BTS is to transmit and receive radio signals from a mobile unit over the air interface. To perform this function effectively, the signals are encoded, encrypted, multiplexed, modulated, and then fed to the antenna system at the cell site. In order to keep the mobile station synchronized, BTS broadcasts frequency and time synchronization signals. Similarly, the received signal from the mobile is decoded, decrypted, and equalized for channel impairments.

Uplink radio channel measurement is made by the BTS and corresponding downlink measurements made by MS.

For the communication between MS and BTS, there are various types of channels to carry the voice & data facilitate call handling & control functions etc. Some channels like Paging, Power control, traffic etc. are dedicated for control functions and traffic channels are dedicated for transportation of user's data (Voice/data).

BTS-BSC Configurations

There are several BTS-BSC configurations: single site; single cell; single site -multicell; and multisite-multicell. These configurations are chosen based on the rural or urban application. These configurations make the GSM system economical since the operation has options to adapt the best layout based on the traffic requirement. Thus, in some sense, system optimization is possible by the proper choice of the configuration.

For example, in rural areas, most BTSs are installed to provide maximum coverage rather than maximum capacity, whereas in urban setup, capacity per BTS is the deciding factor; means in order to cover the same area as in rural more number of BTSs will be required in Urban.

Functions of BSC

The BSC, as discussed, is connected to the MSC on one side and to the BTS on the other. The BSC performs the Radio Resource (RR) management for the cells under its control. It assigns and release frequencies and time slots for all MSs in its own area through the BTSs. The BSC performs the inter cell handover for MSs moving among the BTSs in its control. The BSC controls the power transmission of both BTSs and MSs in its area. The BSC provides the time and frequency synchronization reference signals which is further broadcast by its

BTSs. The BSC also measures the time delay of received MS signals relative to the BTS clock. If the received MS signal is not centered in its assigned timeslot at the BTS, The BSC can direct the BTS to notify the MS to advance the timing such that proper synchronization takes place.

The BSC may also perform traffic concentration to reduce the number of transmission lines from the BSC to its BTSs, or to MSC.

NETWORK SWITCHING SUBSYSTEMS (NSS):

The Network Switching Subsystem includes the main switching functions of GSM as well as the databases needed for subscriber data and mobility management. The MSC also manages the communications between the GSM users and other telecommunication network users. The basic switching function like setting up calls to and from GSM users is performed by the MSC. The MSC has interface with the BSS on one side (through which MSC/VLR is in contact with GSM users) and the external networks on the other (ISDN/PSTN/PSPDN) & other PLMN or MSCs.

Functions of MSC

The main function of the MSC is to coordinate the setup of calls between GSM mobile and other telecom users. Specifically, it performs functions such as paging, resource allocation, location registration, and encryption.

The MSC is a telephony switch that performs all the switching functions for MSs located in a geographical area i.e., in the MSC area. The MSC must also handle different types of numbers and identities related to the same MS and contained in different registers: In general, identities are used in the interface between the MSC and the MS, while numbers are used in the fixed part of the network, such as, for routing. The main difference between a MSC and an exchange in a fixed network is that the MSC has to take into account the impact of the allocation of Radio Resources and the mobile nature of the subscribers and hence has to perform, in addition, the activities required for the location registration and handover.

Specifically, the call-handling function of paging is controlled by MSC. The dynamic allocation of access resources is done in coordination with the BSS. More specifically, the MSC decides when and which types of channels should be assigned to which MS. The channel identity and related radio parameters are the responsibility of the BSS; The MSC also provides the control of interworking with different networks. It is transparent for the subscriber authentication procedure.

The MSC supervises the connection transfer between different BSSs for MSs. This is ensured if the two BSSs are connected to the same MSC but also when they are not. The connection transfer procedure is more complex, when more than one MSCs are involved. The MSC performs billing on calls for all subscribers based in its areas. When the subscriber is roaming elsewhere, the MSC obtains data for the call billing from the visited MSC. The exchange of signaling information on the various interfaces towards the other network elements and the management of the interfaces themselves are all controlled by the MSC. Finally, the MSC serves as a SMS gateway to forward SMS messages from Short Message Service Centers (SMSC) to the subscribers and from the subscribers to the SMSCs. It thus acts as a message mailbox and delivery system.

Visitor Location Register (VLR)

The VLR is co-located with an MSC. A VLR may be in charge of one or several MSC LA's. The VLR constitutes the databases that support the MSC in the storage and retrieval of the data of all subscribers present at given time, in its area. When an MS enters the MSC area borders, it signals its arrival to the MSC, which subsequently stores its identity in the VLR. An MS roaming in an MSC area is controlled by the VLR, responsible for that area. When a MS appears in a LA, it starts a registration procedure. The MSC for that area notices this registration and transfers the identity of the LA where the MS is situated to the current VLR. The information necessary to manage the MS is available in the HLR and is transferred to the VLR so that they can be easily retrieved whenever required.

HOME LOCATION REGISTER

The HLR is a database that permanently stores data related to a given set of subscribers registered to the GSM operator in the area served by that HLR. The HLR is the reference database for subscriber parameters. Various identification numbers and addresses as well as authentication parameters, services subscribed, and special routing information etc. are stored. Current subscriber status including a subscriber's temporary roaming number and associated VLR, if the mobile is roaming, is also maintained in the HLR.

The HLR provides data needed to route calls to all home-based MS-SIMs in its MSC area, even when they are roaming out of area or in other GSM networks. The HLR provides the current location data needed to support searching for and paging the MS-SIM for incoming calls, wherever the MS-SIM may be. The HLR is responsible for storage and provision of SIM authentication and encryption parameters needed by the MSC where the MS-SIM is operating. It obtains these parameters from the AUC.

The HLR maintains records such as which supplementary service each user has subscribed to and provides permission for granting services to the user depending on that. The HLR stores the identification of SMS gateways those have messages for the subscriber until they can be transmitted to the subscriber and receipt is acknowledged.

In HLR some data are mandatory, some others are optional. Both the HLR and the VLR can be implemented in the same equipment in an MSC (co-located). A PLMN may contain one or several HLRs. More than one MSC can be connected to one HLR.

AUTHENTICATION CENTER (AUC)

The AUC stores information that is necessary to protect communication through the air interface against intrusions, to which the mobile is vulnerable. The legitimacy of the subscriber is established through authentication and ciphering, which help to protect the user information against unwanted disclosure. Authentication information and ciphering keys are stored in a database within the AUC, which protects the user information against unwanted disclosure and access. This is achieved by sending some randomly generated secret key words (Random Number). Even while the MS on roaming, cipher key is sent through the visited MSC.

The random number and cipher key is supposed to change with each phone call, so finding them on one call will not benefit using them on the next call.

The HLR is also responsible for the "authentication" of the subscriber each time he makes or receives a call. The AUC, which actually performs this function, is a separate GSM entity that will often be physically installed with the HLR. Being separate, it will use separate processing equipment for the AUC database functions.

EQUIPMENT IDENTIFY REGISTER (EIR)

EIR is a database that stores the IMEI numbers for all registered ME units. The EIR uniquely identifies all the registered MEs. There is generally one EIR per PLMN. It interfaces to the various HLRs in the PLMN. The EIR keeps track of all ME units in the PLMN. It maintains various lists of messages. The database stores the ME identification and has nothing do with subscriber who is receiving or originating call. There are three classes of ME that are stored in the database and each group has different characteristics.

- White List: contains those IMEIs that are known to have been assigned to valid MS's. This is the category of genuine equipment.
- Black List: contains IMEIs of mobiles that have been reported stolen.
- Gray List: contains IMEIs of mobiles that have problems (for example, faulty software and wrong make of the equipment). This list contains all MEs with faults not important enough for barring.

ECHO CANCELER

Echo Canceller is used on the PSTN side of the MSC for all voice call between Mobile subscriber and the landline. The EC is required at the MSC-PSTN interface to reduce the effect of GSM delay when the mobile is connected to the PSTN circuit. GSM link through a hybrid transformer in the circuit. The delay causes the echo, which does not affect the land line subscriber but is an annoying factor to the mobile. The standard EC cancels the delay.

OPERATION AND MAINTENANCE CENTER

Maintenance cover both technical and administrative actions to maintain and correct the system operation, or to restore normal operations after a breakdown, in the shortest possible time. The status of network devices can be checked, and tests and diagnostics on various devices can be invoked through OMC. The OMC also provides alarm-handling functions to report and log alarms generated by the network entities. The maintenance personnel at the OMC can define the criticality of the alarm.

The fault management functions of the OMC allow network devices to be manually or automatically removed from or restored to service. For example, diagnostics may be initiated remotely by the OMC. A mobile call trace facility can also be invoked. The performance management functions include collecting traffic statistics from the GSM network entities and archiving them in disk files or displaying them for analysis. Because of the potential to collect large amounts of data, maintenance personnel can select which of the detailed statistics to be collected based on requirements and past experience. As a result of performance analysis, if necessary, an alarm can be set remotely.

The OMC provides control system for changes for the software revisions and configuration of data bases in the network entities. The OMC also keeps track of the different software versions running on different subsystem of the GSM.

MOBILE INTELLIGENT NETWORK (MOBILE IN)

Mobile IN is used in combination with the public Land Mobile Network(PLMN). It consists of service nodes that provide advanced services to subscribers. Mobile IN functions include the Service Switching Point (SSP) and the Service Control Point (SCP) or a combined Service Switching and control point (SSCP). Mechanisms to support operator-specific services that are not covered by standardized GSM services even while roaming outside the Home PLMN are provided by the Customized Applications for Mobile network Enhanced Logic (CAMEL).

Overview of the GSM Interfaces

The GSM specifications define various interfaces for connecting the various nodes in the GSM network. The GSM interfaces covered in this chapter are as follows:

Air interface or Um –interface

The Air Interface is the interface between the BTS (Base Transceiver Station) and the MS (Mobile Station). The air interface is required for supporting:

— Universal use of any compatible mobile station in a GSM network

— A maximum spectral efficiency

A bis –interface

The A bis -interface is the interface between the BSC (Base Station Controller) and the BTS. The interface comprises traffic and control channels. Functions implemented at the A bis -interfaces are:

— Voice-data traffic exchange

— Signaling exchange between the BSC and the BTS

— Transporting synchronization information from the BSC to the BTS

A-interface

The A-interface is the interface between the BSC and the MSC.

The U m –interface

One of the most important interfaces is the U m or Air interface. This interface is thoroughly specified to achieve a full compatibility between mobile stations of various manufacturers and networks of different operators.

FDMA and TDMA methods

To achieve a high spectral efficiency in the cellular network a combination of FDMA (Frequency Division Multiple Access) and TDMA (Time Division Multiple Access) is used. The FDMA part involves the division by frequency of the 25 MHz bandwidth into 124 carrier frequencies spaced 200 KHz for GSM-900. For GSM-1800 the frequency spectrum of the 75 MHz bandwidth is divided into 374 carrier frequencies spaced 200 KHz. One or more frequencies are assigned to each BTS. Each of these carrier frequencies is then divided in time, using a TDMA scheme to increase the number of channels per carrier frequency.

Each carrier frequency channel carries eight time-division multiplexed physical channels. A physical channel is determined by the carrier frequency (or a number of carrier frequencies and a defined hopping sequence) and the timeslot number. A mobile station can transmit speech data only during its assigned timeslot.

Uplink and downlink

In the frequency range specified for the GSM-900 mobile radio networks, 124 frequency channels with a bandwidth of 200 KHz are available for both the uplink and downlink direction. The uplink (mobile station to BTS) uses the frequencies between 890 MHz and 915 MHz and the downlink (BTS to mobile station) uses the frequencies between 935 MHz and 960 MHz. The duplex spacing, the spacing between the uplink and downlink channel, is 45 MHz.

GSM-1800 uses a similar scheme. The difference is that for GSM-1800 the uplink uses the frequencies between 1710 MHz and 1785 MHz and the downlink the frequencies between 1805 MHz and 1880 MHz. The duplex spacing is 95 MHz.

Types of Logical Channels:

1. Traffic channel $\longrightarrow$ TCH / F
$\longrightarrow$ TCH / H

2. Control Channel $\longrightarrow$ Broadcast channels
$\longrightarrow$ Common control channels
$\longrightarrow$ Dedicated control channels

Logical Channels on the U m –interface

One or more logical channels can be transmitted on a physical channel. There are different types of logical channels. The type of logical channel is determined by the function of the information transmitted over it.

The following types of logical channels are defined:

- Traffic channels
- Broadcast channels
- Common control channels
- Dedicated control channels

Note that the first channel type carries speech and data, and the other types control information (signaling).

Traffic Channels:

The traffic channels are used to send speech or data services. There are two types of traffic channels. They are distinguished by their transmission rates.

The following traffic channels are provided:

- TCH/F (Traffic CHannel Full rate)

    The TCH/F carries information at a gross bit rate of 22.8 kbit/s (after channel coding). The net (or effective) bit rate at the TCH/F is for speech 13 kbit/s and for data 12, 6 or 3.6 kbit/s (before channel coding). The transmission rates of the data services allow services which are compatible to the existing, respectively, 9.6, 4.8 and 2.4 kbit/s PSTN and ISDN services.

- TCH/H (Traffic CHannel Half rate)

    The TCH/H carries information at a gross bit rate of 11.4 kbit/s. The net bit rate at the TCH/H is for speech 5.6 kbit/s and for data 6 or 3.6 kbit/s.

    A TCH/F or a TCH/H may also be used to send signaling information (for example call forwarding and short messages). In that case a small portion of the time slot is used.

A. Broadcast Channels:

- BCCH
- FCCH
- SCH
- CBCH

B. Common Control Channels:

- PCH
- AGCH
- RACH

C. Dedicated Control Channels:

- SACCH
- FACCH
- SDCCH

Broadcast Channels:

The information distributed over the broadcast channels helps the mobile stations to orient themselves in the mobile radio network.

The broadcast channels are point-to-multipoint channels which are only defined for the downlink direction (BTS to the mobile station). They are divided into:

- BCCH (Broadcast Control CHannel)

Via the BCCH the mobile station is informed about the system configuration parameters (for example Local Area Identification, Cell Identity and Neighbor Cells). Using this information, the mobile stations can choose the best cell to attachto.

The BCCH is also known as beacon.

- FCCH (Frequency Correction CHannel)

To communicate with the BTS the mobile station must tune to the BTS. The FCCH transmits a constant frequency shift of the radio frequency carrier that can be used by the mobile station for frequency correction.

- SCH (Synchronization CHannel)

The SCH is used to time synchronize the mobile stations. The data on this channel carries the TDMA frame number and the BSIC (Base Station Identity Code).

- CBCH (Cell Broadcast CHannel)

The CBCH is used for the transmission of generally accessible information (Short Message Service messages) in a cell, which can be polled by the mobile station.

Common Control Channels:

Common control channels are specified as point-to-multipoint channels which only operate in one direction of transmission, either in the uplink or downlink direction. The following channels are provided:

- PCH (Paging CHannel)

The PCH is used in the downlink direction for paging the mobile stations.

- AGCH (Access Grant CHannel)

The AGCH is also used in the downlink direction. A logical channel for a connection is allocated via the AGCH if the mobile station has requested such a channel via the RACH.

- RACH (Random Access CHannel)

The RACH is used in the uplink direction by the mobile stations for requesting a channel for a connection. It is an access channel that uses the slotted Aloha access scheme.

Dedicated Control Channels:

Dedicated control channels are full-duplex, point-to-point channels. They are used for signaling between the BTS and a certain mobile station. They are divided into:

- SACCH (Slow Associated Control Channel)

The SACCH is a duplex channel which is always allocated to a TCH or SDCCH. The SACCH is used for transmission of signaling data, radio link supervision measurements, transmit power control and timing advance data. Note that the SACCH is only used for non-urgent procedures.

- FACCH (Fast Associated Control Channel)

The FACCH is used as a main signaling link for the transmission of signaling data (for example handover commands). It is also required for every call set-up and release. During the call the FACCH data is transmitted over the allocated TCH instead of traffic data; this is marked by a flag called a stealing flag. The process of stealing a TCH for FACCH data is called pre-emption.

- SDCCH (Stand-alone Dedicated Control Channel)

The SDCCH is a duplex, point-to-point channel which is used for signaling in higher layers. It carries all signaling between the BTS and the mobile station when no TCH is allocated. The SDCCHs are used for service requests (for example Short Message Service), location updates, subscriber authentication, ciphering initiation, equipment validation and assignment to a TCH. The net SDCCH bit rate is about 0.8 kbit/s.

Multiplexing Logical Channels on to Physical Channels:

Several of the above-mentioned types of logical channels can be transmitted over one single physical channel (timeslot). The GSM specifications 05.02 specify several combinations of channel types (the sequence of logical channels is fixed).

The order of the logical channels depends on the channel combination.

Channel Combinations:

The channel combinations are:

1. TCH/F + FACCH/F + SACCH/F
2. TCH/H + FACCH/H + SACCH/H
3. (TCH/F + FACCH/F + SACCH/F) or (TCH/H + FACCH/H + SACCH/H)
4. FCCH + SCH + CCCH + BCCH
5. FCCH + SCH + CCCH + BCCH + SDCCH/4 + SACCH/4
6. CCCH + BCCH
7. SDCCH/8 + SACCH/8

Definitions:

The CCCH is a channel that carries both the PCH and the AGCH on the downlink, and the RACH on the uplink.

The extensions "/4" and "/8" in the above-mentioned terms mean, respectively, that four and eight logical channels are mapped onto one physical channel (timeslot).

Note that the BCCH is always transmitted in timeslot 0 on the first defined frequency.

Frame Types on the U m –interface:

The GSM specifications define several types of frames, which are:
- TDMA frame

A TDMA frame consist of eight timeslots (physical channels). The length of a timeslot is 0.577 ms. The length of a TDMA frame is therefore 4.62 ms.

Note: because data on a timeslot is transmitted in bursts, the length of a timeslot is often expressed in BP (Burst Period); 1 BP represents the length of 1 timeslot.

- 26-TDMA multiframe

This multiframe is defined as a succession of 26 TDMA frames, and corresponds to the 26 x 8 BP or 120 ms cycle used in the definition of the TCH/F and TCH/H.

- 51-TDMA multiframe

This multiframe is defined as a succession of 51 TDMA frames, and corresponds to the 51 x 8 BP cycle used in the definition of the TCH/F, TCH/H and of the common channels.

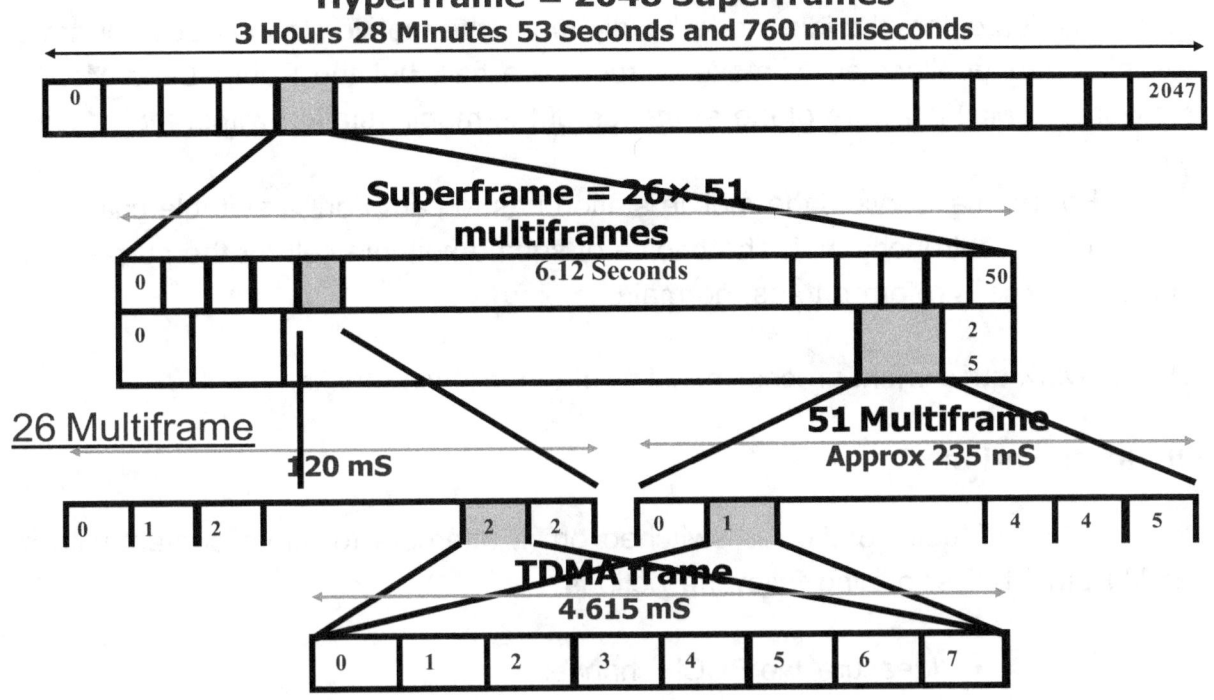

GSM Radio Interface Cycles

- Superframe

    The superframe is a succession of 51 x 26 TDMA frames (6.12 sec), and corresponds to the smallest cycle for which the organization of all channels is repeated.

- Hyperframe

    The hyperframe is the numbering period. It is 2048 x 51 x 26 x 8 BP long, or 3 hours, 28 minutes, 53 seconds and 760 milliseconds. It is a multiple of all previously cited cycles, and determines all the cycles in the transmission of the radio path. It is in particular the smallest cycle for frequency hopping and for ciphering.

Network Attachment Process:

    Network attachment is the process of selecting an appropriate cell (radio frequency) by the mobile station to provide the available services, and making its location known to the network.

The process starts when the mobile is switched on, and ends when the mobile station enters the idle mode. In idle mode the mobile station does not have a traffic channel allocated to make or receive a call, but the Public Land Mobile Network (PLMN) is aware of the existence of the mobile station within the chosen cell.

For normal service, the mobile station has to camp on a suitable call, tune to that cell's control channel. The choice of such a suitable cell for the purpose of normal service is referred to as "normal camping".

The network attachment process consists of the following tasks:

Cell Identification

When Mobile Station is switched on, it attempts to make contact with a GSM PLMN by performing following actions.

- Measure the BCCH channel.
- Search for a suitable call.

The Mobile Station measure the signal strength of the BCCH (Broadcast Control Channel) channels received. It stores a list information about 30 of these BCCH channels, such as signal strength and the frequency corresponding.

PLMN Selection

The particular PLMN to be contacted can be selected either in one of the following modes:

Automatic mode

In automatic mode, the station will choose which PLMNs to try by itself. The automatic mode is based on the existence of the preferred list, which is stored in a non-volatile memory in the SIM. This list includes a number of PLMN identities in order of preference and is under control of the user. The most preferred is usually the home PLMN. The list is filled in by the user through a mechanism to be specified by the mobile station manufacturer. The automatic mode is normally used when the mobile station operates on its home PLMN (that is the PLMN the mobile station is subscribed to).

Manual mode

In manual mode, the user is presented a list containing all found PLMNs. The user chooses one of the PLMNs from the list.

Cell selection:

The mobile station attempts to find a suitable cell by passing through the list in descending order of received signal strength; the first BCCH channel which satisfies a set of requirements. The requirement that a cell must satisfy before a mobile station can be provide service from it, are:

1. It should be a cell of the selected PLMN.
    The mobile station checks whether the cell is part of the selected PLMN.

2. It should not be "barred".
    The PLMN operator may decide not to allow mobile stations to access certain cells. These cells may, for example, only be used for handover traffic. Barred cell information is broadcast on the BCCH to instruct mobile stations not to access these cells.

3. The radio path loss between the mobile station and the selected BTSmust be below a threshold set by the PLMN operator.

4. It should not be in an LA which is in list of "forbidden LA's for roaming".

No suitable cell found

If the mobile station is unable to find a suitable cell to access, it attempts to access a cell irrespective of the PLMN identity, and enters a "limited service" state in which it can only attempt to make emergency calls.

When successfully camped on a cell the mobile station regularly looks to see if there is a better cell in terms of a cell-re-selection criterion, and if there is, the better cell is selected. Also, if one of the other criteria changes (e.g the current serving cell becomes based), or there is downlink signaling failure, a new cell is selected.

CHAPTER-4: CALL PROCESSING IN GSM

In this chapter, aspects of the call processing are explained and looked into specifics case of a mobile originated (MO) call and a mobile terminated (MT) call. Short message (SMS) and voice mail service (VMS) as implemented IMPCS pilot project are also explained.

RF channel overview:

RF channel play important role in call processing case. These are basically three types of RF control channel.

1. Broadcast control channel:

The broadcast channels are points to multi-point channel, which are defined only for down-link direction (BTS to mobile station). They are divided into:

BCCH (Broad cast control channel):

BCCH acts as a beacon. It informs the mobile about system configuration parameters (e.g., LAI, CELL IDENTITY, NEIGHBOURING CELL). Using this information MS choose the best cell to attach to.

BCCH is always transmitted on full power and it is never frequency hopped.

FCCH frequency correction channel:

MS must tune to FCCH to listen to BCCH. FCCH transmits a constant frequency shift of the radio carrier that is used by the MS for frequency correction.

SCH (synchronization channel):

SCH is used to synchronize the MS in time. SCH carries TDMA frame number and BSIC (Base Station Identity Code)

2. Common control channels:

Common control channels are specified as point to multi-point, which operate only in one direction either in up-link or down-link direction.

PCH (Paging Channel):

PCH is used in down-link direction for sending paging message to MS whenever there is incoming call.

RACH (Random Access Channel) :

RACH is used by the MS to request allocation of a specific dedicated control channel (SDCCH) either in response to a paging message or for call origination /registration from the MS. this is an up-link channel and operate in point to point mode.

AGCH (Access Grant Channel):

AGCH is a logical control channel which is used to allocated a specific dedicated control channel (SDCCH) to MS when MS request for a channel over RACH. AGCH is used in downlink direction.

3. Dedicated Control Channel:

Dedicated control channel is full duplex, point to point channel. They are usedfor signaling between the BTS and certain MS. They are divided into: -

SACCH (Slow Associated Control Channel):
the SACCH is a duplex channel, which is always allocated to TCH or SDCCH. The SACCH is used for

- Radio link supervision measurements.
- Power control.
- Timing advance information.

In 26 frame traffic multi-frame 13th frame (frame no .12) is used for SACCH.SACCH is used only for non-urgent procedures.

FACCH (Fast Associated Control Channel):
FACCH is requested in case the requirement of signaling is urgent and signaling requirement cannot be met by SACCH. This is the case when hand-over is required during conversation phase. During the call FACCH data is transmitted over allocated TCH instead of traffic data. This is marked by a flag known as stealing flag.

SDCCH (Stand Alone Dedicated Control Channel):
The SDCCH is a duplex, point to point channel which is used for signaling in higher layer. It carries all the signaling between BTS & MS when no TCH is allocated to MS. The SDCCH is used for service request, location updates, subscriber authentication, ciphering. equipment validation and assignment of a TCH.

Mobile originated (MO) call:

There are four distinct phase of a mobile originated call-

-Setup phase.
-Ringing phase.
-Conversation phase.
-Release phase.

Out of these phases the setup phase is the most important phase and includes authentication of the subscriber, Ciphering of data over radio interface, validation of mobile equipment, validation of subscriber data at VLR for requests service and assignment of a voice channel on A-interface by MSC. Whenever MS wants to initiate on outgoing call or want to send an SMS it requested for a channel to BSS over RACH. On receiving request from MS, BSS assigns a stand-alone dedicated control channel (SDCCH) to MS over access grant channel (AGCH). Once a SDCCH has been allocated to MS all the call set up information flow takes place over SDCCH.

A connection management (CM) entity initiates a <u>CM Service Request</u> message to the network. Network tries to establish an MM connection between the MS and the network and upon successful establishment of MM connection a <u>CM Service Accept</u> message is received by MS from the network. MS now sends a <u>Call Set up Request</u> to the network which contains the dialed digits (DD) of the called party. As the call setup message is received at the MSC/VLR certain check are performed at MSC/VLR like- whether the requested service is provisioned for the subscriber or not, whether the dialed digits are sufficient or not, any operator determined barring (ODB) does not allow call to proceed further etc. As these checks are performed at MSC/VLR a Call Proceeding Message is sent from the network towards the MS. After all the checks are successfully passed MSC sends Assignment command to the BSS which contains a free voice channel on A-interface on getting this message BSS allocates a free TCH to the MS and informs the MS to attach to it. MS on attaching to this TCH informs the BSS about it. On receiving a response from the BSS, MSC switches the speech path toward the calling MS. Thus, at the end of Assignment the speech path is through from MSto MSC. It is important to note that at this stage mobile has not connected user connection as yet. MS at this stage does not listen anything.

After assignment MSC sends a network set-up message to the PSTN requesting that a call be set up. Included in the message are the MS dialed digits (DD) and details specifying which trunk should be used for the call. The PSTN may involve several switching exchanges before finally reaching the final local exchange responsible for applying the ringing tone to the destination phone. The local exchange will generate the ringing tone over the trunk, or series of trunk (if several intermediate switching exchanges are involved), to the MSC. At this point in time MS will hear ringing tone. The PSTN notifies the MSC with a network-alerting message when this event occurs. MSC informs the MS that the destination number is being alerted. It is important to note that this is primarily a status message to the MS. The MS hears the ringing tone from the destination local exchange through the established voice path.

When the destination party goes off hook, PSTN informs the MSC of this event. At this point, MS is connected to the destination party and billing is started. MSC informs the MS that connection has been established and MS acknowledges the receipts of the connect message.

Under normal condition, the termination of a call is MS initiated or network initiated. In this scenario, we have assumed that MS initiates the release of the call by pressing "end" button and MS send a disconnect message to the MSC. The PSTN party is notified of the termination of the call by a release message from the MSC. The end- to- end connection is terminated. When MSC is left with no side task (e.g. charging indication etc.) to complete a release message is sent to the MS. MS acknowledges with a release complete message. All the resources between MSC and the MS are released completely.

Mobile Terminated (MT) call:

The different phases of a mobile terminated call are

- Routing analysis
- Paging.
- Call setup.
- Call release.

The phases of mobile terminated (MT) call are similar to a mobile originated (MO) call except routing analysis and paging phase. Call to a mobile subscriber in a PLMN first comes to gateway MSC (GMSC). GMSC is the MSC, which is the capable of querying HLR for subscriber routing information. GMSC need not to be part of home PLMN, though it is normal practice to have GMSC as part of PLMN in commercially deployed networks.

GMSC opens a MAP (Mobile Application Part) dialogue towards HLR and Send / Routing / Info-Request (SRI request) specific service message is sent to HLR. SRI request contains MSISDN of the subscriber. HLR based on location information of this subscriber in its database, opens a MAP dialogue towards VLR and sends Provide / Roaming / Number-request (PRN request) to the VLR. VLR responds to PRN request with PRN response message, which carries an MSRN (mobile subscriber roaming number), which can be used for routing toward visiting MSC in the network. HLR returns MSRN to GMSC (MSC that queried HLR) in SRI response message. On getting MSRN the GMSC routes the call towards VMSC The purpose of this entire exercise is to locate where the terminating mobile subscriber is.

The MSRN received at GMSC is in international format (Country Code + Area Code + subscriber number). Normally, based on the routing info at GMSC, the call may be routed out of the GMSC towards VMSC of the terminating subscriber, in which case appropriate signaling protocol (MF or ISUP) depending on the nature of connecting of GMSC with subsequent exchange along the route will apply. If at VMSC the terminating mobile subscriber is found to be free (idle), paging is initiated for terminating mobile subscriber. MSC uses the LAI provided by the VLR to determine which BSS's should page the MS. MSC transmit a message to each of these BSS requesting that a page be performed. Included in the message is the TMSI of the MS. Each of the BSS's broadcasts the TMSI of the mobile in a page message on paging channel (PCH).

When MS detects its TMSI broadcast on the paging channel, it responds with a channel request message over Random-Access Channel (RACH). Once BSS receives a channel request message, it allocates a stand –alone Dedicated Control Channel (SDCCH) and forwards this channel assignment information to the MS over Access Grant Channel (AGCH). It is over this SDCCH that the MS communicates with the BSS and MSC until a traffic channel assigned to the MS. MS transmits paging response message to the BSS over the SDCCH. Included in this message is MS TMSI and LAI. BSS forwards this paging response message to the MSC. Now Authentication and Ciphering phases are performed to check the authenticity of MS and encrypt data over radio interface.

On the network side after paging is initiated, while waiting for paging response, a defensive timer called," Early ACM" timer is run at MSC to avoid network timeouts. On successfully getting paging response, a setup message is constructed to be sent towards terminating MS. In case paging fails due to authentication failure or when the subscriber is out of radio-coverage, the call is cleared.

In case CLIP is not subscribed by the terminating mobile subscriber, calling number is not included in set-up message. In case CLIP is subscribed and PI value in calling number parameter indicates "presentation allowed" the number is included in the set-up message. In case CLIP is subscribed but PI received in calling number parameter indicates "presentation restricted" then number is included only if CLIRO is also subscribed to.

MS on receiving the set-up message performs compatibility Checking before responding to the set-up message – it is possible that MS might be incompatible for certain types of call set-ups. Assuming that MS passes compatibility checking, it acknowledges the call setup with set-up confirm message. After getting set-up confirm message from the MS, MSC performs assignment phase (similar to one discussed in MO call) and a voice path is established from MSC to the MS. MS begins altering the user after it receives the traffic channel assignment. MS send alerting message to the MSC .MSC upon receiving the alerting indication from the MS, begins generating an audible ringing tone to the calling party and sends a network alerting via GMSC to the PSTN. Prior to this the calling party heard silence.

At this point in the call, MS is alerting the called party by generating on audible tone. One of the three events can occur-calling party hangs-up, mobile subscriber answers the phone, or the MSC times out waiting for the mobile subscriber to the answer the call. Since radio traffic channel is a valuable resource, GSM does not allow a MS to ring forever.

In the present scenario we have assumed that the mobile subscriber answers the phone. The MS in response to this action stops alerting and sends a connect message to the MSC. MSC removes the audible tone to the PSTN and connects the PSTN trunk to BSS trunk (terrestrial channel) and sends a connect message via GMSC to the PSTN. The caller and the called party now have a complete talk path. This event typically marks the beginning of the call for billing purposes. MSC sends a connect acknowledge message to the MS.

The release triggered by the land user is done in similar way as the release triggered by mobile user. MSC receives a release message from the network to terminate end-to-end connection. PSTN stops billing the calling landline subscriber. MSC sends a disconnect message towards the MS and MS responds by a Release message. MSC release the connection to the PSTN and acknowledges by sending a Release Complete message to PSTN. Now the voice trunk between MSC and BSS is cleared, traffic channel (TCH) is released and the resources are completely released.

The mobile-to-mobile call scenario is a combination of phases encountered in mobile originated (MO) and mobile terminated (MT) call.

Short Message Service (SMS)

SMS is a simple bearer service that functions as a bi-directional alphanumeric paging service, allowing for the provision of value-added services as well as management services such as charge advice. A short message can only have 160 characters (it can be less depending upon the type of characters and their coding scheme). The SMS could be sent in either broadcast or point-to-point mode (via the CBCH channel) (via either SDCCH channel if mobile is in idle state, or SACCH if the mobile is in dedicated mode).

SMS allows to provide many values added service to individual/ corporate clients. Individuals may be interested in messaging (transmitting messages in compact way) or leisure services (weather forecast, road traffic, restaurant booking, movies, TV programs etc.). Business users may be interested in corporate information (company performance, stock value), e-commerce etc.

SMS involves specific entities in the GSM network: first is the SMS Service Centre (SMS-SC simply SMSC) which can be connected to several networks and many MSC's (SMS- GMSC's or SMS-IWMSC's) within the same PLMN and which is addressed by a mobile using a E.164 number of the numbering plan of the PLMN. SMSC is capable of following functionality's:

- Transmission of short message towards a mobile, retaining the responsibility of the message until reception of acknowledgement or expiration of the validity period.

- Reception   of the short messages from MS and transmission ofacknowledgement to the PLMN.

- Transferring messages received from Internet to mobile.

The second entity involved by the SMS is the SME (short message entity), which is responsible for producing or receiving a short message. The SME can be connected to the SMSC via a data network such as X.25 or IP.

A short message is characterized by its parameters the most significant are the validity period, the service center time stamp which indicates the SM arrival time at the SC, etc.

In IMPCS (pilot project), the SMS architecture has been implemented by C-DoT. The hardware architecture of SMSC is similar to HLR and is located on same physical platform. It works as an inter-working and relaying function of the message transfer between two MS. The service provided are-

(i)   Mobile Originated short message - Enables MS to send an SMS (up-to 140 bytes) to another MS via SMSC.

(ii)   Mobile terminated short message - Enables delivery of an SMS to a particular MS.

(iii)   Operator initiated SMS - This facility enables fixed network subscriber to send an SMS to a mobile subscriber through an operator at SMSC.

(iv) SMS Newsletter Service - A group of mobile subscribers can subscribe to SMSC for receiving periodic news regarding sports, weather, traffic etc. Thesubscription is done through on operator at SMSC. The operator feeds the news segments, which are transferred, to the subscriber periodically.

Voice Mail System (VMS)

In the system, VMS provides the function of a call answering device. It provides subscribers with a personal voice mailbox. VMS redirects/forwards voice calls from a subscriber who is temporarily unavailable (busy or no reply) to a personal mailbox of the subscriber connected to the MSC. When a call is redirected to VMS, it first greets the caller with a personalized greeting message and prompts the caller to press a button. Please leave the message in the mailbox. Later, the called party (mobile subscriber) can use an access code to access the VMS from a PLMN/PSTN phone.

VMS communicates with MSC over E1 lines via the R2 MF/CCS#7 signaling protocol. The VMS in the IMPCS network is a Pentium PC equipped with a Dialogic card and running Windows NT 4.0. Dialogic cards support telephony network interfaces, voice recording, compression, and playback. The disk capacity required by the PC is entirely dependent on the application. If each subscriber stores 10 minutes of voice data, then disk storage for subscriber voice information is approximately 20 GB for 10,000 subscribers.